D0432178

BEAR

"EACH AND EVERY TIME
I HAVE MET A BEAR
IN THE WILDERNESS,
I HAVE LEARNED
SOMETHING VALUABLE."

—PAUL NICKLEN

BEAR

SPIRIT OF THE WILD

PAUL NICKLEN

Washington, D.C.

CONTENTS

Svalbard, Norway ~ A curious bear inspects my tiny winter cabin in the snowy landscape *(opposite).*

Admiralty Inlet, Nunavut, Canada ~ A polar bear shakes off excess water after a dive under the sea ice. *(page 2)*

Great Bear Rainforest, British Columbia ~ In the moss-draped rain forest, a spirit bear eats a pink salmon to fatten up for hibernation. *(pages 4–5)*

Fishing Branch River, Yukon Territory ~ A young grizzly explodes into action as it chases a chum salmon. *(pages 6–7)*

Svalbard, Norway ~ The massive landscape of a fjord in Svalbard, Norway, dwarfs a lone polar bear. More than 90 percent of Svalbard's glaciers are receding. *(pages 8–9)*

Great Bear Rainforest, British Columbia ~ A black bear patiently waits for a meal in a fast-moving salmon river. *(pages 10–11)*

Great Bear Rainforest, British Columbia ~ A spirit bear gorges on Pacific crab apples, an important nutritional source for bears. *(pages 12–13)*

FOREWORD

Through time, bears have become iconic figures that embody mankind's relationship to wilderness; they symbolize a footprint through which many cultures, particularly the North American Native, construe life, nature, and spirituality. These magnificent creatures inspire admiration, awe, and tenderness; at the same time they instill fear and respect. Bears are social and playful, as well as introspective and reclusive. Generally, they live peacefully but will not hesitate to use their strength if needed.

For 700,000 years, bears and hominids have shared the Earth, and an intertwined melody between the two families has developed. Many native cultures consider bears healers whose powers are miraculous, shamanic, and magical. For the Dakota, the mere dreaming of a bear is prophetic of a medicinal gift; the Cheyenne call medicinal plants "bear root"; the Winnebago's bear dance is a healing dance; the Zuni use bear claws to treat wounds. Huron, Kootenai, and Navajo often invoke the bear as healer in their dances, prayers, stories, and songs.

Today, bears face an ecological challenge in a competing world between the urban and the wild. They require large and healthy habitats not only to survive and thrive but also to be true to their nature. Bears are an integral part of the lost wisdom of the North, which values all forms of life. We must return full cycle to that understanding; we must stop destroying wilderness and domesticating wildlife for our use and benefit. The belief that human life is above other life has unearthed a roar of protest and a call to action. As said so eloquently in the Lakota prayer Mitakuye Oyasin: "We are all related, my relatives, without whom I would not live. We are in the circle of life together, co-existing, co-dependent, co-creating our destiny."

Since the beginning of civilization, bears have played a role as keepers of our spirit. In listening to the wisdom of these magnificent mammals we can learn the importance of living in unison with nature—to roam and gather its fruits, hunt and fish its riches, take care of our own, retreat into introspection, heal ourselves, and be reborn. In rekindling our respect and reverence for bears, we can only hope to understand and embrace the message that all life is to be treated with respect, honor, and love.

Richard Sneider, Ph.D.
Chairman's Council, Conservation International
Board Member and Trustee, Greater Los Angeles Zoo Association

Admiralty Inlet, Nunavut

After an hour under the ice, I emerge numb from the cold and am humbled by the experience.

PHOTO CREDIT: JED WEINGARTEN

preceding pages

Fishing Branch River, Yukon Territory

A grizzly bear rubs itself on a marking tree near the river. *(page 16)*

INTRODUCTION

PAUL NICKLEN

If you have picked up this book hoping to read about a near-death experience with a bear, you will be deeply disappointed. As you will witness through the images and the stories from these great authors, none of us has a terrifying story to tell; instead, we have all been greatly inspired by the last true nomads of North America.

The genesis for this book started 30 years ago, when I first became enchanted with bears and all they represent. It seems as if my entire career, first as a polar biologist and then as a nature photographer, has followed the trails traveled by bears. From my first experiences with polar bears as a young man in northern Canada to the hundreds of chance meetings I have had with bears in remote corners of the forest, I have always strived to be as nonintrusive as possible—a fly on the wall just watching the bears doing bear things. Each and every time I have met a bear in the wilderness, I have learned something valuable. The more time I spend observing their behavior and trying to understand their role as top predators, the more their role as guardians of a fragile and intact ecosystem becomes evident.

Oftentimes, as I sit quietly on the sea ice watching a polar bear hunt for seals or rest by a remote forest stream spying on a grizzly as it chases salmon, I am reminded of the great passions bears elicit from humans; the predators both fascinate us and terrify us. They are one of the most popular animals in the world today, but to most of us, the thought of encountering a wild bear evokes a sense of terror. Perhaps it's because the media has so successfully portrayed them as fierce, bloodthirsty predators or perhaps it's because

Taking a break from shooting aerials in Alaska, I'm greeted by swarms of mosquitoes.

PHOTO CREDITS *(left to right):* PETER MATHER, JED WEINGARTEN, PETER MATHER, PETER MATHER

A spirit bear, unperturbed by my presence, feeds in the Great Bear Rainforest.

we have a fear of becoming part of the food chain. Whatever the reason, bears are often perceived as dangerous and unpredictable. Nevertheless, at the zoo, crowds marvel and delight as a new cub takes its first steps. Children's books and cartoon characters portray bears as lovable, friendly animals.

Even so, there are those who would happily exploit and feed this fear of bears with sensationalistic stories. Recently, I walked into a bookstore looking for a book about bears, and every title I found read like a horror story: *A Kodiak Bear Mauling: Living and Dying with Alaska's Bears*, *Fighting for Your Life: Man-Eater Bears*, and even the oddly named *Bear Attacks: Their Causes and Avoidance*. Is it any wonder people are terrified of bears when they are portrayed as human-eaters?

Nothing could be further from the truth. Bears are not only incredibly smart and sensitive, they are also superb communicators. In the many years I have spent with them, I have learned that they are also very honest. Bears always show you who they are, and if you take the time to sit quietly with them, you will come out of the experience enriched and gifted with a sense of what is pure and wild. Over the many years I have spent wandering the northern landscapes in search of images, I have had thousands of encounters with polar bears and hundreds of chance meetings with grizzly bears, and I have spent hundreds of peaceful moments

On board my inflatable boat, I power up a small Yukon river in search of grizzlies.

A curious black bear had inspected my tent while I was away taking photos.

in the company of black and spirit bears. I can honestly say I have never felt under threat, and I have never seen a bear act unpredictably or aggressively. This is why I have made it my life's mission to use my images and my passion to give a voice to animals like bears.

Yes, there are rare cases when bears attack humans, but for the most part they are peaceful, gentle, and smart creatures. They simply are misunderstood and unfairly characterized. I want my images to serve as ambassadors for these creatures so that we can find a more harmonious relationship with them. As Phil Timpany so eloquently writes in his chapter, "Nothing good will ever come from killing a grizzly bear."

In the end, I wanted to do this book not only because I have such deep respect for bears but also because of what they represent. Bears are symbols for wild nature, and as apex predators they require vast tracts of wilderness to roam and thrive. Indeed, the presence of bears in an ecosystem is a very good indicator of the overall health of that system. Where grizzly and black bears roam, there are also healthy streams, functioning forests, communities of small and large mammals, birds and reptiles, and the pulse of a full array of ecological processes that allow life to exist on this planet. Where healthy polar bears reign supreme, we also find healthy sea ice and a diverse assortment of species, like ringed seals, belugas, narwhals, and other animals that support populations of polar bears. As apex predators,

I am enveloped by an intense display of aurora borealis painting the night sky.

A grizzly strolls past me in the Yukon Territory's Fishing Branch River.

PHOTO CREDITS *(left to right)*: PETER MATHER, PHIL TIMPANY, JED WEINGARTEN, NATHAN WILLIAMSON

bears are the powerful force that allows ecosystems to thrive. And yet, as wilderness continues to shrink, we are leaving no place for bears to go, and as bears come into contact with humans, a sad story of conflict, slaughter, and fear has developed. Deforestation, melting sea ice, and other human-induced impacts on these ecosystems will eventually lead to the disappearance of bears.

For the past 20,000 years, humans have coevolved and coexisted harmoniously with bears. Bears represent our most constant wild companions as they share most of the same habitats where humans dwell. It has just been with the advent of guns that we have declared war on bears. In order to find a forward path that allows people and bears to coexist in harmony, we need to make it explicit that our fate is connected to the fate of the natural world and of creatures like bears. We need a story that connects to our emotions and helps shift our collective identity toward one of hope and accountability. We need a story in which we are both spectators and key players. That is the story this book is telling, one in which we are all held accountable to the natural ecosystems we depend on and to the creatures that live in them.

Through the essays in this book we can begin to realize that bear problems are people problems. When bears run out of habitat, it only means we too are running out of habitat. As

As I drift out to sea, I push my komatik, or sled, to safer ice.

A black bear, preoccupied with eating a salmon, allows me to get up close and personal.

the last wilderness areas of our planet, the places where bears roam, continue to disappear, so does the potential for humans to enjoy the ecosystem services that wilderness provides.

The water we drink, the food we eat, the fibers that clothe us, the fuels that keep us warm and cool, and the medicines that restore our health are all produced by the intricate web of life. The inspiration for our arts, cultures, and religions—our recreation and aesthetics too—have come from nature. Bears are the living representation of these ecosystems, and their fate is tied to ours.

I have had the privilege of walking down many bear trails in the company of these magnificent animals. I make no claims of special friendships or spiritual connections but will forever be grateful to the bears that have tolerated my presence, allowed me to photograph them, and shared with me a small part of their greatness.

A GLOBAL PERSPECTIVE

A view of the top of the Earth reveals the Northern Hemisphere and displays the distribution of the four species and subspecies of bears featured in this book's photographs—the American black bear and its white counterpart, the Kermode or spirit bear; the polar bear; and the grizzly, or brown bear. The habitats of these bears overlap at times, and the map highlights those areas where two or more species coexist as neighbors, sharing their homes with one another.

Circular diagrams (lower right) illustrate the estimated population for each of these bears—as well as the relative size of the populations to each other. The American black bear by far dominates, with an estimated population of some 900,000. By contrast, the spirit bear, which lives only in coastal British Columbia, numbers a mere 500.

The map also features the minimum extent of Arctic sea ice as measured in 2012. Arctic sea ice—a fragile layer of ice that floats atop the sea—grows naturally during the dark polar winters and retreats when the sun appears in the spring. According to satellite data analyzed by NASA, the 2012 minimum is the smallest area of sea ice recorded to date. The minimum extent, usually reached in September, has been decreasing for decades as Arctic ocean and air temperatures continue to rise. This warming trend imperils the survival of the polar bear and the entire polar ecosystem.

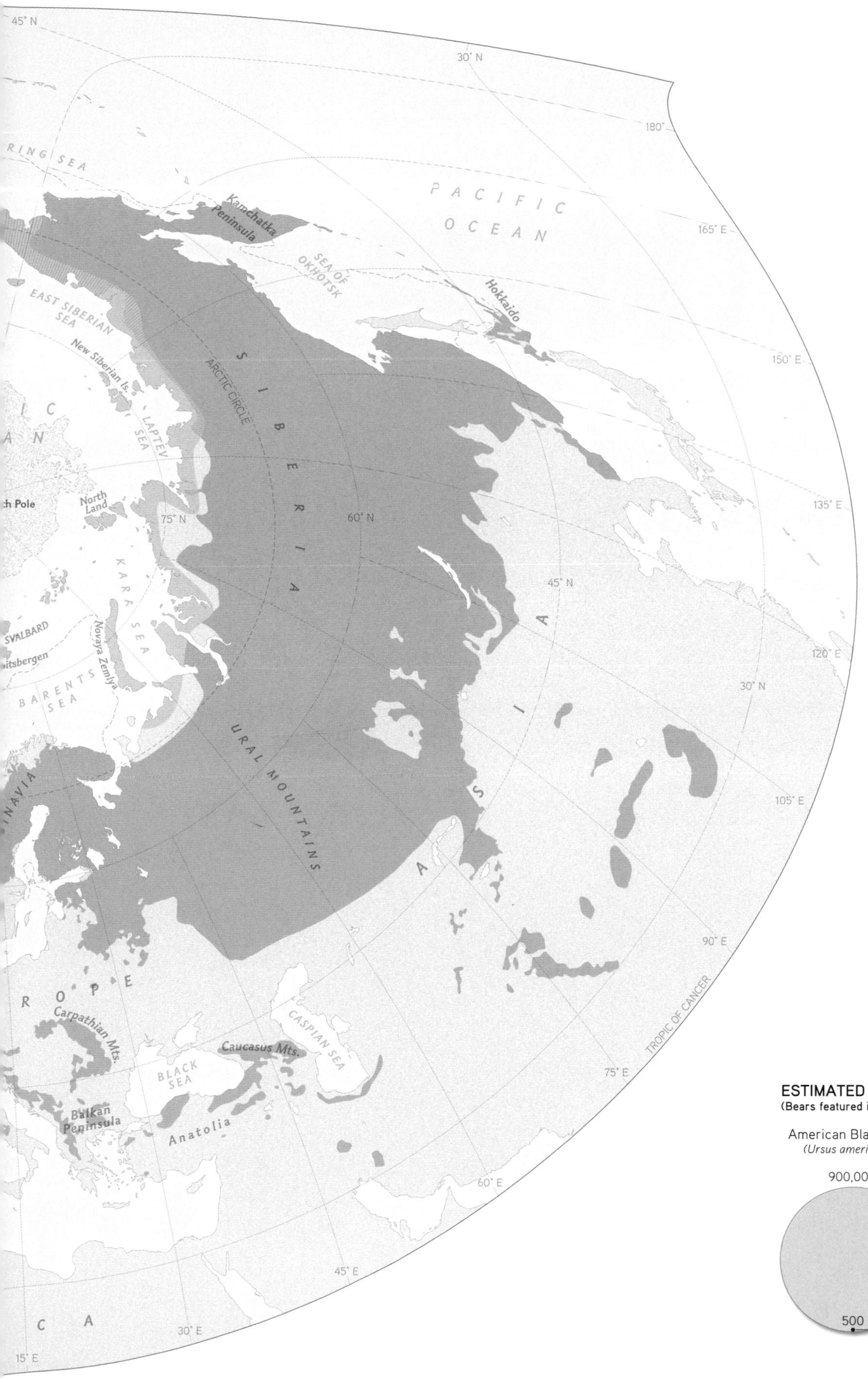

ESTIMATED POPULATION OF BEARS

(Bears featured in this book)

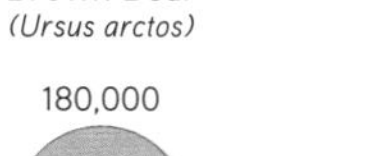

"THE POLAR REGIONS . . . ARE DISAPPEARING IN FRONT OF MY EYES, AND I WANT MY IMAGES TO BECOME BANNERS OF HOPE, AMBASSADORS FOR A WORLD VERY FEW OF US WILL EVER SEE."

—PAUL NICKLEN

AMBASSADORS FOR POLAR WILDLIFE

PAUL NICKLEN

It had been a long time since I had any feeling in my feet or hands as I sat on the sea ice at 22° below zero Fahrenheit. I wanted to jump around, stomp my feet, and swing my arms to entice the feeling back into my frozen limbs, but I did not dare move for fear of scaring away the pair of polar bears I had been watching just 50 feet away.

For 15 years I had been hoping to witness this moment. I have spent countless hours observing polar bears hunting, catching seals, nursing cubs, and sparring with each other. I probably have seen more than a thousand polar bears in the wild, and yet I had never watched them mate. For more than 24 hours I sat quietly on the frozen Barents Sea in Svalbard, Norway, watching the drama of a mating pair that was completely oblivious, yet seemingly aware of my presence. The huge male tried everything in his power to entice the female to mate, thus ensuring that his dominant genes would carry forward into the population.

When we first came upon this pair, my assistant Karl Erik and I parked our snowmobiles facing away from the bears, which allowed us to have an easy escape route in the event one might like to investigate. As we waited in silence, hour after hour, for the couple to consummate their courtship, the vision of capturing this moment I had dreamed about for so many years became a lump in my throat and I had to control my excitement in anticipation.

The thin, weak, and unpredictable conditions of the sea ice surrounding Svalbard in recent years had forced me to put this assignment on hold for several years. Traditionally, sea ice, which gives us access to polar bears in their natural habitat, envelops Svalbard and the surrounding islands nearly year-round. But like everywhere else in the Arctic, the situation is changing fast. Svalbard is now facing its lowest sea ice coverage since scientists began measuring ice thickness and surface area coverage due to warming temperatures. Even though it was March, when the sea ice should be at its largest seasonal extent, we were only able to find a tiny strip of ice on the east side of Svalbard. The bears were plentiful, as there was nowhere else for them to hunt seals. And the lack of ice and opportunities for the bears to hunt meant we had frequent visits to our camp by hungry animals on a near nightly basis. We saw many instances in which bear trails disappeared into the slushy soup of an ailing frozen sea to find solid ice again only several hundred yards away.

We now have irrefutable evidence that the polar regions have warmed at least twice as much as anywhere else on the planet. Scientists inform us that the original projections of the summer extent of polar ice disappearing within the next 100 years are inaccurate. In fact, the Arctic could be completely free of ice during the summer months within the next five to ten years—a terrifying prospect for all Arctic species. Multiyear ice, ice that survives for several years providing the foundation of the polar food chain, will be gone. As one scientist bluntly puts it, "If we lose sea ice, we stand to lose an entire ecosystem."

A close look at sea ice reveals its life-giving properties. After the long winter months, when the sun finally returns to the Arctic, phytoplankton starts to grow on the underside of the ice. The tiny plant communities that thrive there create an inverted garden that becomes the base for the entire food chain. Tiny crustaceans, such as amphipods and copepods, feed on the phytoplankton. The mighty bowhead whale skips all nutritional levels, feeding directly on the copepods and amphipods. Also feeding on this soup of tiny creatures is the Arctic cod—a key component of the Arctic ecosystem, which has a direct effect in up to 75 percent of the energy transfer between the plankton and larger vertebrate species like fish, seals, walruses, and marine birds. The beluga whales and narwhals also feed on the cod. And at the top of the food chain, the apex predator, the polar bear, feeds on seals and walruses. In the absence of sea ice, polar bears lose their hunting platform to hunt seals and are confined to land, a foodless prison where they slowly starve. In Antarctica, the cast of characters is different but leopard seals, crabeater seals, and penguins are caught in the same downward spiral as their Arctic counterparts.

A couple of years ago, I guided former U.S. President Jimmy Carter in a small boat around a herd of walruses in Svalbard. As we looked into the polar realm, he told me that if people are to care about complex and distant issues, like climate change and the loss of sea ice, we need to find a common and simple language to share and spread the facts with the rest of the world. To me, that language is photography, the opportunity to create evocative images that speak for polar wildlife and its dependence on healthy polar environments. The polar regions that have been my lifelong playground are disappearing in front of my eyes, and I want my images to become banners of hope, ambassadors for a world very few of us will ever see.

As dire as things are, polar bears are thriving in many areas, and I still love finding big males hunting or females playing with their cubs or young males sparring. On that cold spring day when I found the mating pair, I waited for as long as I could. Just when I thought I could not take the cold any longer, the female finally got up, calmly walked over to the now sleepy male, and nudged him to mate. I wiped my frosty breath from the viewfinder, ready to capture this rarely seen behavior. At that exact moment, another large male appeared out of the corner of my eye. He ran at the other male, roaring. The female ran away onto the rotten sea ice with the males in hot pursuit. I was left sitting on the frozen ground, in complete disbelief. The moment was over. I had lost the chance of a lifetime, but will never forget being so present in the drama of nature.

A polar bear dives under the ice and glassy calm surface of Nunavut's Admiralty Inlet, casting its reflection in the Canadian Arctic water *(opposite).*

preceding pages

A polar bear raises its snout as it picks up my scent during a blizzard at Churchill, Manitoba. *(page 26)*

The windswept sea ice reveals the tracks of a polar bear making its way toward the snowy terrain of Svalbard, Norway. *(pages 28–29)*

A polar bear cautiously navigates the disintegrating ice pack off the northeast corner of Spitsbergen, Norway. *(pages 34–35)*

From the vantage point of my ultralight airplane, I photographed a pod of male narwhals in Lancaster Sound, Nunavut. As they come up for air, narwhals can become a target for polar bears.

A young polar bear leaps between ice floes on the Barents Sea near Svalbard, Norway.

A ringed seal emerges from the water of Canada's Admiralty Inlet, scanning for polar bears before taking a breath.

A skinny polar bear eats a beluga whale after other bears had consumed their fill near North Spitsbergen, Norway *(above).*

A large male polar bear feeds on the ribs of a bearded seal in Leifdefjorden, Norway *(opposite).* He was so full that he was regurgitating out of one side of his mouth while continuing to eat with the other.

following pages
A large male polar bear travels across the sea ice near Churchill, Manitoba, in anticipation of a winter of feasting on seals.

Polar bears do mock battle with each other in a wintry blizzard on Hudson Bay, Manitoba *(opposite)*. This is practice for when they battle for real in the spring when they fight for the right to breed.

Rich Arctic light illuminates a bear at 2 a.m. near Ellesmere Island, Nunavut, as it strolls across the sea ice *(above)*. Powerful moments like this make me reflect and wonder if bears will exist in a hundred years, given their shrinking habitat.

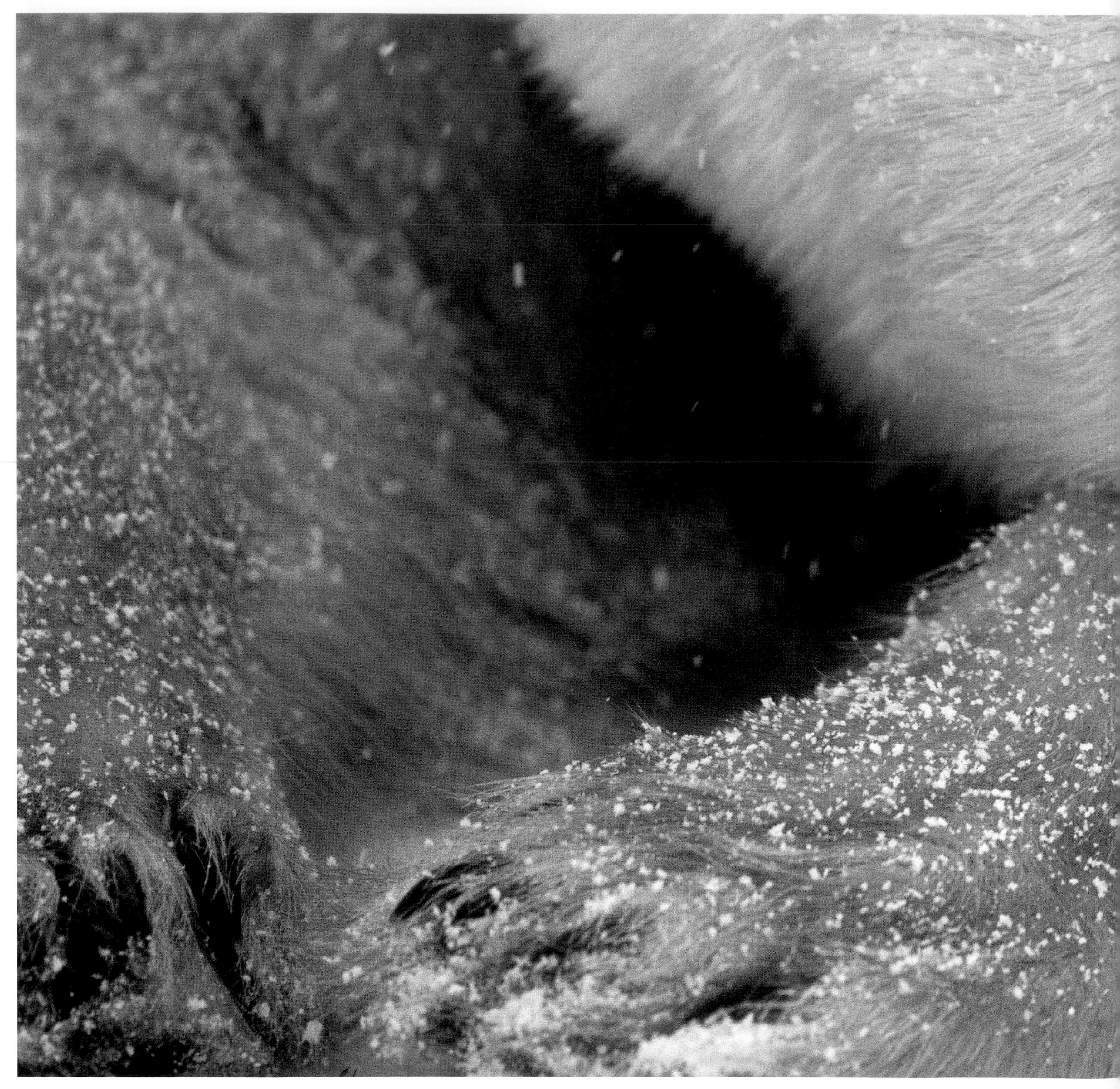

A male polar bear unsuccessfully tries to entice a female to mate near East Svalbard, Norway *(opposite)*. For hours, I patiently waited in the frigid cold to capture this rarely observed behavior.

preceding pages

To conserve energy, this bear waits out a blizzard near Churchill, Manitoba. Watching polar bears has taught me incredible patience. *(pages 50–51)*

A fogbow glows over Foxe Basin, Nunavut, at 1 a.m., highlighting the melting multiyear ice. *(pages 52–53)*

following pages

A protective mother scans the Manitoba landscape for danger from male polar bears.

Near Hudson Bay, Manitoba, a female polar bear nurses her cub *(above)*, while another prepares to feed her two hungry youngsters *(opposite)*. When stranded on land, mothers are under great nutritional pressure to feed their cubs as the mothers lose weight.

following pages

A mother and her cub travel across the sea ice in the glow of the low winter sun of Hudson Bay, Manitoba.

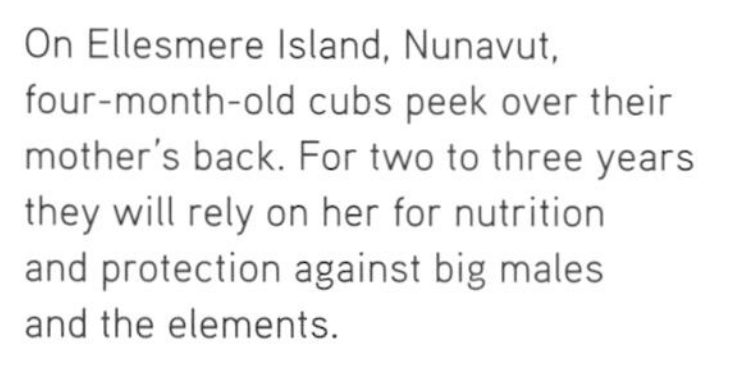

On Ellesmere Island, Nunavut, four-month-old cubs peek over their mother's back. For two to three years they will rely on her for nutrition and protection against big males and the elements.

A female polar bear and her two cubs, just seven months old, stand along the basalt shores of Spitsbergen, Norway. Once stranded on land, the bears have little access to food until the ice freezes again in the late fall.

An Arctic fox sniffs for leftovers near the shores of Hudson Bay, Manitoba *(opposite)*. An adult fox weighs just eight pounds, so any scraps from an unsuspecting polar bear go a long way during the long winter months.

Remains of a polar bear sprawl across the sea ice on Nunavut's Lancaster Sound *(above)*. Finding dead bears is a rare occurrence but is becoming more commonplace. I have even found bears floating dead in the open ocean near Svalbard, Norway.

A battle-wounded male polar bear hunts at minus 40°F near Ellesmere Island, Nunavut. Wounds like this are common and they heal easily.

A polar bear amuses itself by balancing a block of snow on its forehead near Hudson Bay, Manitoba *(above and opposite)*. When I came upon the scene, I was eager to create a sequence of images to capture the humor of the moment.

following pages

A large pan of multiyear sea ice melts in Alaska's Beaufort Sea. One of the most important challenges I have undertaken is communicating how warming temperatures are contributing to the loss of ice in polar regions.

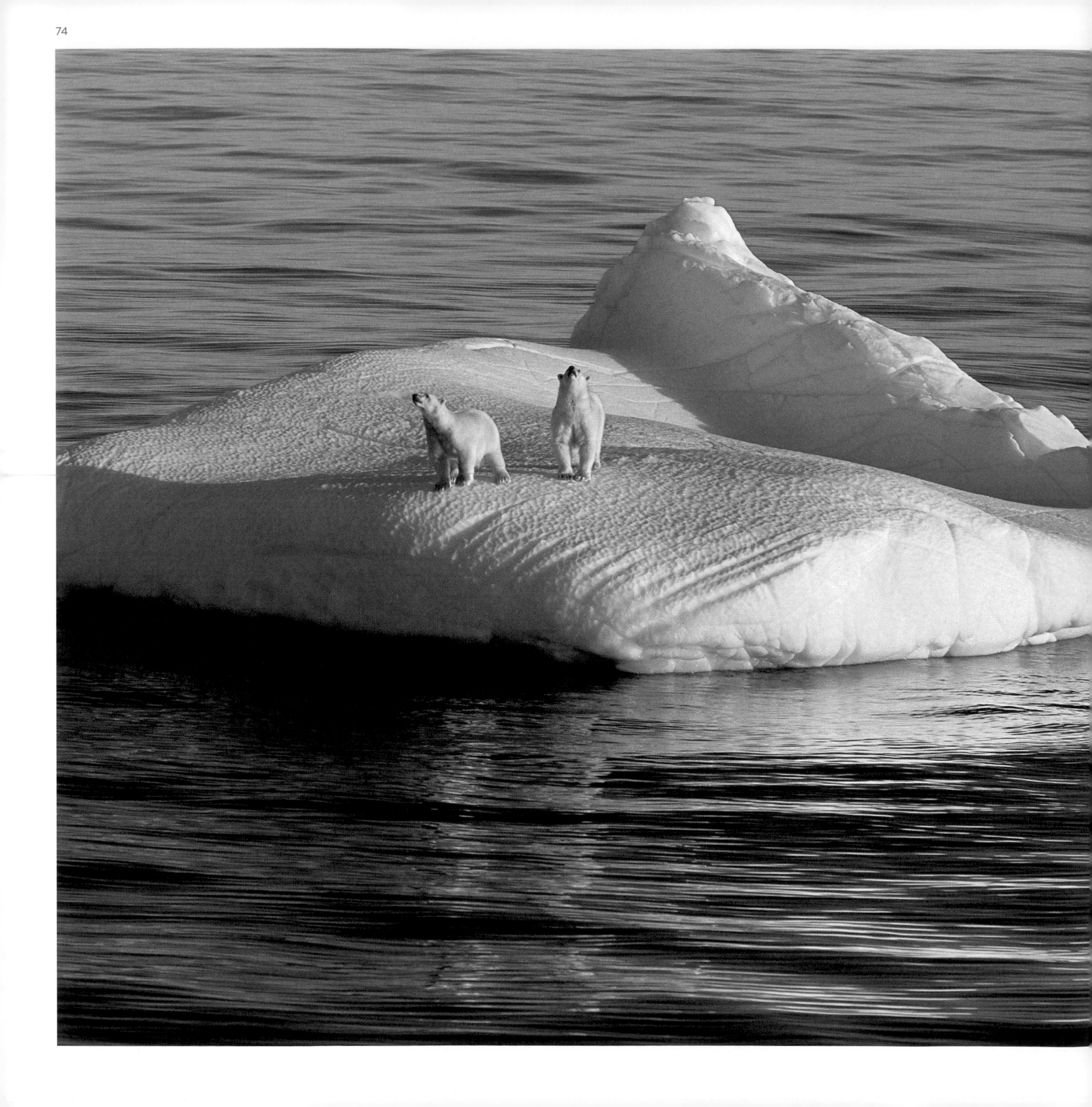

A PERSONAL PERSPECTIVE

ICON OF THE ARCTIC

SYLVIA EARLE

Standing on the frozen lip of one of the many islands of Franz Josef Land, I followed the fluid movement of a lone white figure making its way across a patch of floating ice and watched as it slipped into the sea as naturally as a bird takes to the sky. En route to the North Pole on a Russian icebreaker, I was enjoying a brief break from our ocean voyage to the frozen top of the world. It was my first glimpse of the ultimate icon of the Arctic, a wild polar bear. The stark landscape covering the Arctic Ocean seems an improbable home for polar bears, warm-blooded carnivores that may weigh half a ton. In a realm of alabaster white, the secret to the success of bears in the Arctic lies a few feet under their paws, beneath the ice, in the sea below.

As our ship hammered its way through pack ice, room-size chunks flipped over, revealing rich gardens of life. Surface temperature may plunge below a life-numbing minus 58°F and winds may howl, but the ocean can never have a temperature below a relatively benign 28°F. Golden fields of microscopic diatoms and sweeping plains of glowing green algae and cyanobacteria cling like paint to the underside of the ocean's frozen ceiling, some drifting free into the liquid blue below, where the average depth is more than 3,250 feet. Minuscule cells gather the summer sun's light, generate oxygen, take up carbon dioxide, reproduce themselves, and help power hordes of copepods, isopods, amphipods, and other invertebrates no larger than the commas on this page. This miniature zoo gorges on the captured sunlight, and in turn is consumed by small jellyfish, larval fish, arrow worms, and other tiny predators. Here, life abounds with abundance and diversity, culminating with predatory fish, birds, seals, whales—and one very special kind of seagoing bear, *Ursus maritimus.*

To make just one pound of polar bear food requires many tons of microscopic algae that in turn have given life to a long and twisted web of diners who take up energy—and give back nutrients that in turn keep the cycle moving. From plankton to fish to the bear's preferred menu of bearded and ringed seals, the heightened flow of energy in summer carries through the continuous era of winter darkness. The system has worked for millions of years of bear history, and in the past few thousand years for tribes of people who arrived and shared the Arctic's sharp seasonal rhythms.

My most recent encounter with a polar bear was in a zoo in China where a captive-born baby swam in a refrigerated pool glassed on one side so crowds of adoring children could see the bear's dark eyes and ebony nose, and how in the water its fur fluffed into a silken robe. For the children, it was a moment of joy; for me, a time of ineffable longing that today's children might grow into a world where wild polar bears are respected and treasured as a critical part of an ancient system vital to the way Earth functions, a system now under siege owing to human activity. The bears cannot know why or what to do about what is happening to their once pristine realm, but we know, and we can take measures to protect the wild Arctic, a place vital for their future—and ours.

A mother polar bear and her two-year-old cub are stranded on glacier ice in Canada's Hudson Strait.

"NOTHING GOOD
WILL EVER COME
FROM KILLING
A GRIZZLY BEAR.
MUCH GOOD CAN COME
FROM RESPECTING ITS
RIGHT TO CONTINUE
TO ROAM THE LAND."

—PHIL TIMPANY

TWO THOUSAND DAYS WITH GRIZZLY BEARS

PHIL TIMPANY

"Spectacular," said the Belgian, looking through his camera as we sat near a stream in northwestern Canada. Twice the grizzly bear had lunged into the water to catch a chum salmon and brought it to shore to eat in peace. I looked at the bear, at the Belgian, then again at the bear. The animal was indeed spectacular. It consumed the fish and started walking toward us.

"Is it close enough?" he asked. I whispered, "Yes." A bolt-action rifle—a weapon with a short bolt designed to prevent jamming—replaced the camera. He placed the crosshairs on the bear and began to shake. I said, "Take your time." He readjusted the rifle and took aim. The shaking started again. At 150 feet, the bear turned broadside to study a group of spawning chum. "Breathe," I whispered, "and put the bullet low, just behind the front shoulder." The rifle roared. Water erupted at the bear's front feet. The animal spun around and bellowed, waving a nearly severed and bloody front paw. "Shoot again," I urged. The Belgian tried putting another round in the chamber. It didn't work. His rifle had jammed.

The grizzly began running across the slough to get away. I picked up my rifle and shot. The bullet entered the bear and exited in a large spray of blood and lung tissue. It was a quick death. For me, this was anything but spectacular. The Belgian, however, was happy. He had paid $8,000 to slay the bear.

Thirty-seven years later, I sit by the Fishing Branch River a few miles south of the Arctic Circle with a small group of tourists. We are watching a 500-pound male grizzly chase salmon. My task for the morning has been successful. I have safely placed people and a grizzly together in a natural setting. My 12-gauge shotgun is leaning against my pack, loaded and ready for use. The bear-viewing business I share with the Vuntut Gwitchin First Nation of Old Crow requires that I carry it. In the 21 years I have been working here, I have never fired it. Morris, the 15-year-old male, walks ten feet from us, undisturbed by the shutters of the cameras. Guests have been watching him fish on this river since 2006. He has starred in natural history documentaries, and his photos have been published globally in many magazines.

As a young man, I dreamed of a career with wildlife. I thought I could fulfill my naturalist ambitions by working as a guide for “sportsmen” who came from around the world to kill iconic wildlife, especially grizzlies. It pains me to confess that I am responsible for the deaths of many grizzlies. My appreciation for these magnificent mammals, and an increasing repulsion for what I came to see as senseless killing, ended my guiding career. With the bloody adventure behind me and my desire for wilderness still intact, I was lucky to find a vocation that allowed me to begin a long relationship with a population of grizzly bears. In the mid-1970s, I began doing fieldwork for a scientific study of Chinook salmon in a remote area of northern British

Columbia frequented by grizzlies. To date, I've spent more than 2,000 days with grizzly bears. It has been a privilege, a life lesson, and a humbling experience. I am deeply touched by their intelligence, physical power, forgiving nature, and honesty. The familiarity I have developed with these bears now allows me to share my experience with tourists and photographers in wilderness areas in northern British Columbia and the Yukon Territory. When we enter the bear's domain, we engage in a peaceful coexistence based on respect.

Grizzlies are captivating animals and most people believe they are worthy of protection. Opposing views, however, exist as to how to go about it. The institutions supporting the recreational killing of grizzly bears justify the activity on economic terms. They profess their followers have great love for the animal and the killing gives the bear a value, which encourages its protection. The moral issues surrounding the slaughter never enter the equation. Others, like me, believe in the full protection of grizzly bears. We feel that maintaining viable populations of this apex predator and preserving a strong genetic pool will secure the vitality of the entire ecosystem. The morality of killing grizzly bears is an integral part of this thinking. The time I have spent in the presence of grizzly bears has molded my philosophy, making me a staunch advocate for their protection. I see no justifiable reason or need to kill them.

The animal's harmonious and spiritual relationship with indigenous peoples has been replaced by one in which the great bear is now a target for those who wish to kill it, study it, manage it, photograph it, or simply retain a memory of it. Where I live and work in the Yukon and British Columbia, "wildlife management" philosophies are, for the most part, archaic artifacts of early colonial enterprise. They are 100 years old in their framework of legislation and, in regard to grizzly bear management, at least 50 years behind public opinion. As the rest of the world increasingly recognizes the value of apex predators, my government continues to squander this wildlife resource by refusing to create a responsible vision that fulfills its economic and ecological potential. The politics surrounding grizzly bears involves making money, providing job security, and giving those with a passion for killing these animals the legal and moral means to do so.

For now, the best argument for conserving grizzly bears continues to be economic—although my hope is that people will adopt economy-based values that do not include killing them. Over the long term, the survival of grizzly bears will only be guaranteed if they are granted some form of citizenship within our society complete with their habitat requirements. Their continued existence will have to be as important as our own. Some of us will have to learn to coexist with them, suffer with them, and on rare occasions die by them. There is still time for governments and citizens to reframe their attitudes and relationships with these great bears. Without a new culture of appreciation and tolerance, the grizzly bear will vanish.

Nothing good will ever come from killing a grizzly bear. Much good can come from respecting its right to continue to roam the land.

A grizzly leaps across a pond. On my first day on the Fishing Branch River, I went from hoping to see a bear to having this sub-adult appear just ten feet away from me *(opposite)*.

preceding pages

A cautious young female surveys her surroundings in the male-dominated Fishing Branch River Valley. *(page 76)*

The Yukon's vast Peel Watershed is home to a healthy population of bears but remains under threat due to recent gold discoveries and trophy hunting. *(pages 78–79)*

A young female grizzly makes her way to the Fishing Branch River. Every morning I waited for this bear to appear out of the frost-covered forest, almost like a ghost. *(pages 84–85)*

On the lookout for large males, a young bear approaches the Yukon Territory's Fishing Branch River to eat. The river is a stressful location for young bears as they compete with older bears for food and space.

On many days, bears would visit guide Phil Timpany's camp at Bear Cave Mountain on the Fishing Branch River and take their turn marking the territory with a good rub against a tree *(above and opposite)*.

following pages

A bear wades through the Fishing Branch River in late autumn. I cherish the transition between seasons in the polar ecosystems as fall gives way to the long, dark, cold winter along the Arctic Circle.

After weeks of waiting, a large male finally inspects my underwater housing in the Fishing Branch River. Even though I was sitting just 20 feet away, all of his attention was focused on the silver housing.

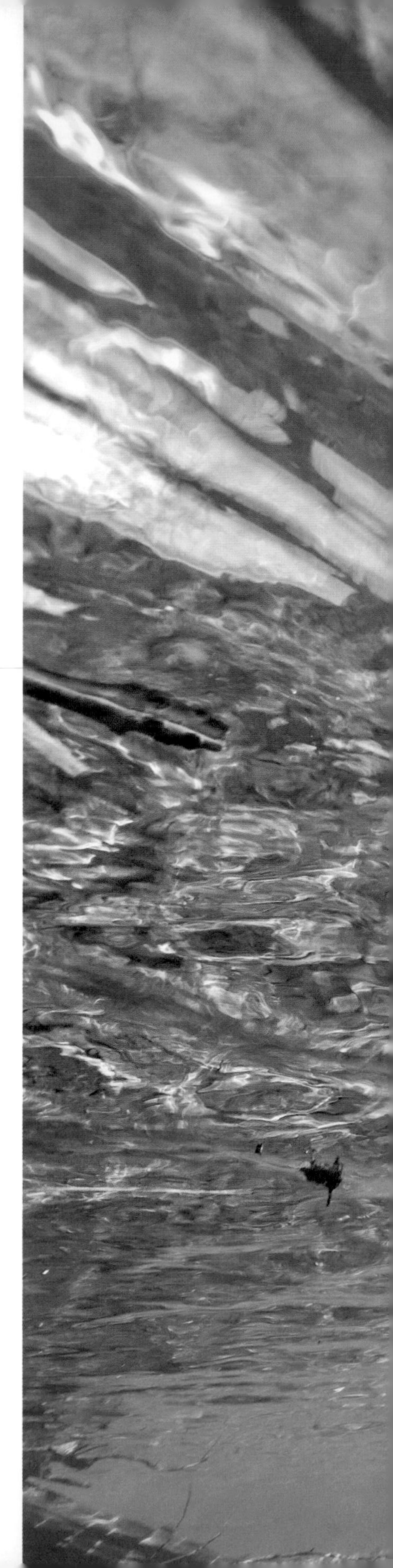

A dead salmon is caught in a back eddy of the Fishing Branch River, the perfect location to find an easy meal *(above)*.

A bear claws at my underwater housing in the Fishing Branch River. The bear could not resist inspecting the underwater dome. Within seconds, his curiosity was satisfied and he never came near it again *(opposite)*.

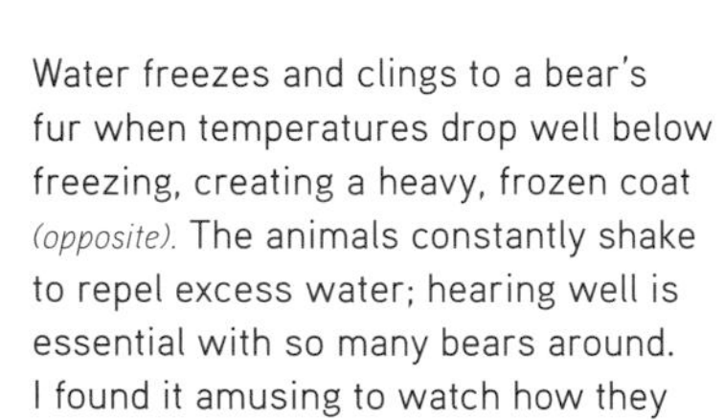

Water freezes and clings to a bear's fur when temperatures drop well below freezing, creating a heavy, frozen coat *(opposite)*. The animals constantly shake to repel excess water; hearing well is essential with so many bears around. I found it amusing to watch how they hated getting their ears wet.

following pages

A young bear wades in the Fishing Branch River, after struggling to catch salmon. This vantage point gave us many great viewing opportunities every day.

A cold night leaves the willows along the Fishing Branch River covered in hoarfrost—a sign that the bears will soon enter their dens for the winter *(opposite)*.

preceding pages

With the red fall colors in full swing, the aurora borealis lights up the night sky and the Yukon's rugged landscape at 3 a.m.

following pages

A young bear is on full alert as it approaches a dying salmon along the Fishing Branch River's edge.

Thousands of caribou migrate through the North Slope of Alaska's Arctic National Wildlife Refuge en route to the Yukon Territory for the winter. Barren-ground grizzlies and wolves follow the caribou until the beginning of the salmon and berry seasons.

The Fishing Branch River is a place of tension for the bears, but they completely ignored us. The large males *(above)* were often displacing the younger, shy bears *(opposite)*.

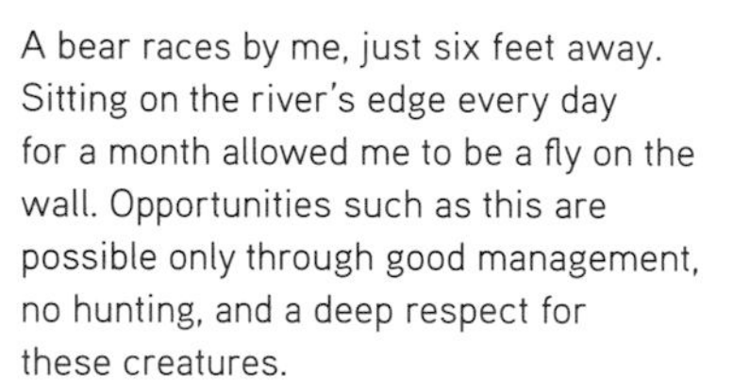

A bear races by me, just six feet away. Sitting on the river's edge every day for a month allowed me to be a fly on the wall. Opportunities such as this are possible only through good management, no hunting, and a deep respect for these creatures.

A young bear tries with no success to nab a salmon in the Fishing Branch River *(above and opposite)*. The younger the bear, the more often it fails in its attempts to catch a fish, and the better the photographic opportunities.

following pages

After several attempts, the young bear's pounce finally results in success, catching one of many fish that day.

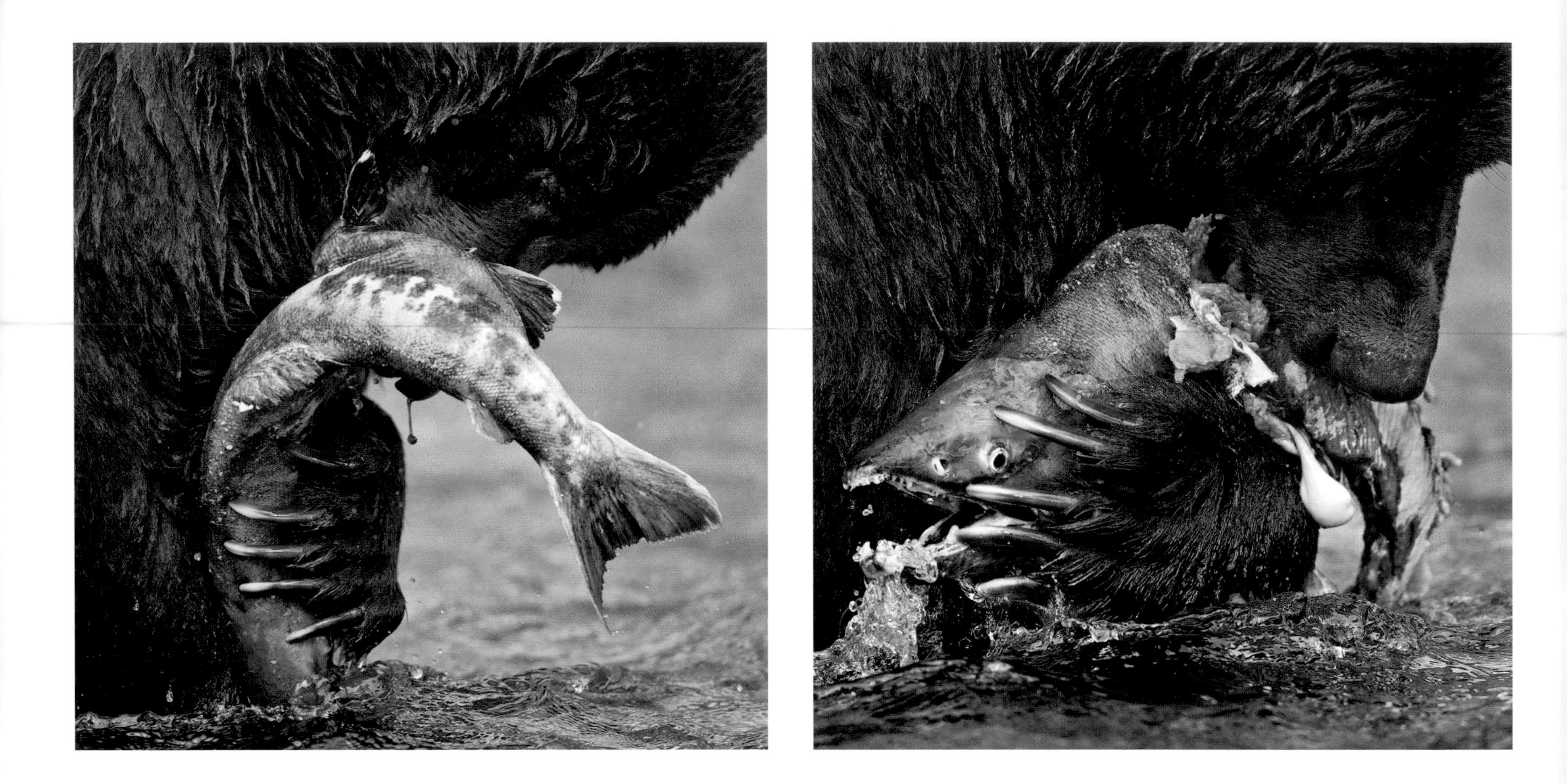

A rotting old chum salmon provides life-sustaining nutrition for a large male bear. The skin and eggs are often their preferred bounty.

I watched this young male bear stick his face under the river's surface looking for salmon, wait for a few seconds, then quickly lift his head to assess the surroundings for safety *(opposite)*.

following pages

Flying back from a month of photographing bears on the Fishing Brach River, I saw this deep blue lake in Tombstone Territorial Park—a reminder that we must protect entire ecosystems where bears roam, not just river systems. With recent discoveries of gold in the Yukon, many layers of the ecosystem may be threatened.

A PERSONAL PERSPECTIVE

ENCOUNTERING GRIZZLIES

WERNER HERZOG

My first encounter with a grizzly bear was a right of passage of sorts for my son. He and I traveled to the Alaska wilderness on a fishing trip to celebrate the end of his childhood. One day, while fishing, we saw a big male grizzly bear on the opposite bank of a wide, torrential river. Although the bear was aware of us, he seemed largely uninterested. I remember how we marveled at his fishing skills as he deftly chased and caught salmon after salmon. We didn't worry too much when he swam across the river and came right at us, or rather at the two sockeye salmon we had caught. Within seconds, he was nearly upon us. I whistled, I yelled, and I shook my arms. He stopped abruptly and stood up on his hind legs—to our amazement, he was as tall as a basketball hoop. I feared for what might happen next, but when I shouted at him, he fled through the swamp, splashing water high up into the air.

Several years later, while I was doing the film *Grizzly Man*, which followed in the footsteps of self-proclaimed bear activist Timothy Treadwell, I had a chance to spend much more time with many grizzlies. Their beauty and power has never ceased to amaze me. Smart and sensitive, they reign undisputed over the ecosystem in which they live. Their reputation for being ferocious is well deserved, but only when the big males fight each other. Harm to humans, however, is rarely seen. Within 100 years of records in Alaska, fewer than 20 fatalities have been recorded. Statistically, bees and wasps kill at least 100 times more humans every year.

Media hysteria over bear maulings might be at fault for this undeserved reputation. Timothy Treadwell, for example, crossed a line when, in trying to protect the bears, he thought of himself as a bear. Timothy was an unusual individual. He would drop on all fours, trying to behave like a grizzly. He saw them with an almost childlike love. "Good day, beautiful bear. You look like an angel. I love you," he wrote about one encounter. He would speak to grizzlies in a high-pitched child's voice; he sang to them and stroked their face. It ended catastrophically when he and his girlfriend were attacked and eaten by a bear. Timothy failed to recognize bears as wild creatures, and in the end he did a disservice to the efforts to conserve them. Incidentally, the bear that attacked him was also killed.

Perhaps, when it comes to bears, it is best to listen to ancient knowledge. The native Alutiiq people tell us that we should never *love* the bear; we should rather *respect* him and keep our distance. They also admonish us that life without such majestic creatures like the mountain lion, the eagle, and the bear would leave us humans impoverished. I agree with them.

This powerful image evokes the words of Phil Timpany: "Nothing good will ever come from killing a grizzly bear."

"AS A FATHER,
I TAUGHT MY GIRLS
NOT TO FEAR
BLACK BEARS,
BUT TO
UNDERSTAND THEM."

—WADE DAVIS

BLACK BEAR COUNTRY

WADE DAVIS

In British Columbia, we live in black bear country, even in the heart of the city. Vancouver lies at the mouth of the Fraser River, at the base of the Coast Mountains. When the salmon run, bears come out of the hills to feast along neighborhood creeks. One such morning, a yearling strolled up my sister's driveway just as the babysitter, a young woman from the Philippines, was taking the kids to the local park. The nanny was unperturbed. She thought it was just another strange and very large Canadian dog.

Most of my bear encounters have been more comical than perilous. When I was 16 and working on a government trail crew in the Canadian Rockies, a black bear slashed the wall of our mess tent and ate an entire summer's supply of powdered Jell-O. In the morning, we could track his discomfort by the small pools of psychedelic scat that colored the trail all the way across the divide to the lakes on the far side of the park. Some years later, while employed as a guide in the Spatsizi, British Columbia's largest wilderness park, I returned to our spike camp to learn that a black bear had tried to enter our kitchen tent at midday. Our cook, less fearful than indignant, simply whacked it on the nose with a frying pan and sent it scurrying on its way.

One time at the headwaters of the Taku River, at the beginning of a three-week rafting expedition, a sow got into our wine. By morning, she was long past drunk, no doubt suffering from a wicked hangover. Nothing we did could get her and her cub to move on. For the obligatory pre-trip safety talk, we had the clients face the river so that a guide with a shotgun could discreetly patrol the edge of the forest behind them. The expedition leader was demonstrating the proper way of getting into a raft when the sow feigned a charge. In a heartbeat, our entire mob plunged into the rafts, headfirst with legs flying, as those on the oars pulled hard for the far shore.

Conventional wisdom says that each bear needs 10,000 acres of unbroken forest to be content, and that the range of a single animal stretches for ten miles or more. You would be hard-pressed to get my two daughters to believe this. Growing up, they spent every summer at our fishing lodge at Ealue Lake, a thousand miles north of Vancouver. Ours is the only private holding on a pristine lake six miles long.

On any summer night, Tara and Raina would count 30 or more bears grazing along the shoulder of the right-of-way as we drove north on the Stewart Cassiar, the one road running through a region the size of Oregon. On the dirt track that spurs seven miles off the highway to our lodge, they always tallied piles of scat. Their record, as I recall, was 127. In the dry and dense poplar thickets that rise above our lake, bear sign is everywhere: tracks, scat, claw marks on the trees, fireweed and hellebore crushed in small beds in the meadows. From a young age, Tara and Raina led guests up a game trail to the tundra and summit of Sky Mountain, or Eya-dzitla—a faint track through the bush that intimidates many an adult client. Encountering a bear, most especially at the end of the day in the spruce forest that skirts the base of the mountain, was not uncommon.

As a father, I taught my girls not to fear black bears, but to understand them. Should you cross paths with a bear, stand up to it, never run, face it directly, and talk to it softly, even as you back away, never breaking eye contact. If the bear snorts, snaps its teeth, blusters into a charge, stand your ground. It's bluffing. A killer bear comes at you slowly, cautiously, in a quiet stalk, ready for the kill. Forget about climbing a tree. With long curved claws, a black bear can scramble up anything. As for hiding, don't even think about it. As the saying goes, if a pine needle falls in a forest, an eagle will see it, a deer will hear it, and a black bear will smell it. The most effective defense is to anticipate and respect a bear's needs. The only dangerous encounters are those with mothers protecting cubs or old toothless bears parasitized to the point of starvation.

I always told Tara and Raina that encounters with black bears are not to be feared, but rather embraced as gifts of the wild. These beautiful animals are the only species of bear to have evolved in the Americas. For 400,000 years they roamed this hemisphere alone, before being joined by their upstart cousins, grizzlies and polar bears, which crossed from Asia a mere 100,000 years ago. Unlike their showy rivals, black bears like people, and there are times, native elders say, when messages pass between the bears and us.

At five, my younger daughter, Raina, disappeared from our property at Ealue Lake. I searched for several hours before finding her a mile or more up the creek, wearing her favorite purple bonnet and work gloves, as she knelt over a dead black bear cub, trying in vain to stroke it back to life. I immediately looked around for the boar that had no doubt killed the cub and now quite possibly threatened my child. Seeing nothing, I turned back to Raina, whose face ran with tears. After a few quiet moments, we laid the cub on a bed of moss, and made our way home. Living with us at the time was Alex Jack, a Gitxsan elder and legendary guide who had lived all his life in the bush as a hunter and trapper. His native name was Axtiigeenix, "he who walks leaving no tracks." When I told Alex what had happened, he took it as a sign, of what I was never sure.

Some years later, Raina suffered a near fatal illness. After she had stabilized, Tara and I accepted an invitation to attend a Navajo peyote ceremony in New Mexico. While I was in the sweat lodge with the elders, Tara worked with the women preparing the feast. At some point she described her sister's condition to the roadman's wife, who immediately shared the information with her husband. Later that night in the tepee, as the fire burned and shadows flickered across the thunderbird altar, he turned to me and said out of the blue, "It was that black bear cub." I was astonished, for Tara had said nothing of her sister's encounter in the woods with the dead cub. But that night the roadman turned all of his healing energy toward Raina, with chants and prayers that lasted well into the dawn. I have no idea what transpired, but I can say without hesitation that it marked the beginning of her recovery, a rebirth that happily is now complete.

After sitting for months in the constant rain of British Columbia's Great Bear Rainforest, I was delighted to see golden rich light reflect off the thick dark fur of a black bear *(opposite)*.

preceding pages

A huge male black bear reigns supreme on a tiny salmon-filled stream in the Great Bear Rainforest. *(page 126)*

An aerial reveals the rugged terrain of coastal British Columbia. The first time I flew over the immense and beautiful area, I wondered how I would ever find black and spirit bears in this vast habitat. *(pages 128–129)*

In early fall, bears like this large male pull themselves away from the berry- and crab-apple-covered mountains to gorge on salmon-filled streams of the rain forest. *(pages 134–135)*

Two large males—a black bear and a spirit bear—stare each other down for the best fishing hole *(above)*. Moments later, they lock in a roaring embrace of claws and teeth as they battle for the best fishing grounds *(opposite)*. In certain areas of the Great Bear Rainforest, as many as one in ten black bears are white.

A black bear makes one of the season's first catches of pink salmon in the rain forest river *(above)*. As the bears fill up, they focus more on the skin and eggs of the fish.

This huge male always knew that I was sitting just above him along the river *(opposite)*. Every once in a while, he would glance up to make sure I had not moved. Out of respect for his patterns, I always stayed in one place so he knew I was not a threat. He and others rewarded my efforts by not leaving the river when I was present.

Black bears were always plentiful on these small creeks in the Great Bear Rainforest. The animals kept even spacing between one another so as to avoid conflict. The most dominant bears had the preferred fishing spots.

Drenched by rain, a black bear shakes and creates an explosion of water droplets while never taking its eye off its quarry.

The black bears each had their own style of catching fish in the rain forest rivers *(opposite).* This large male found ways to rest and yet remain ready to pounce at the sight of a fish.

following pages

Pink salmon have a two-year life cycle. We timed our expedition into the Great Bear Rainforest during the season when the rivers were boiling with salmon.

Bears target female salmon that are full of eggs *(above)*. In the Great Bear Rainforest, the bears ate other salmon quickly, but slowed down to savor a fish with eggs.

A black bear peruses the remnants of a mostly consumed salmon *(opposite)*. When photographing life and death in the animal kingdom, I sometimes need to remove my own emotions from the situation and focus on making pictures.

A camera trap catches a black bear crossing a river in the Great Bear Rainforest *(opposite).* The trap sat in this location for an entire month and took only one image of a black bear and one image of a spirit bear.

following pages

With the ever increasing demand for timber, much of the Great Bear Rainforest has been logged. Flying over this scene reminded me of what a pristine habitat should look like.

A PERSONAL PERSPECTIVE

THRIVING ON THE BORDER

FABIAN OBERFELD

Long before borders divided North America and undocumented migrants risked their lives to cross those borders, black bears roamed freely across the western regions of North America. *Ursus americanus,* commonly known as the black bear, once wandered in great numbers across Canada, the western region of the United States, and in the eastern and western Sierra Madre mountain ranges of northern Mexico.

Despite being highly adaptable, black bears prefer green, forested areas, as these provide abundant food. Yet as forests have changed over time, black bears have learned to survive in many different habitats, including those made by man, like suburban areas. The destruction and fragmentation of forests in northern Mexico in favor of agriculture and urbanization as well as the persistent hunting of these great bears have reduced their population significantly. From an astonishing two million bears that once roamed the land, it is estimated that in all of North America the black bear population is down to 900,000, mostly in Canada and the United States. In Mexico, where low densities make it almost impossible to find this elusive animal, it is hard to know how many survive today.

Dating back thousands of years, the black bear has been an important icon in Amerindian mythology. The Tarahumara, or Rarámuri Indians, who have lived in the western parts of the Sierra Madre mountain range in Chihuahua since time immemorial, know the black bear as *ojui,* a character that appears in some of their stories about the creation of the world. They thought that the *hikuli,* or peyote, a hallucinogenic cactus used in certain rituals, was stronger than the ojui. They smoked peyote to protect themselves against these strong animals. In a story told by the Tarahumara to Carl Lumholtz, a Norwegian ethnographer who studied their rituals, the bear told the hikuli: "'Let's smoke and fight immediately after.' They smoked and fought, and the hikuli was stronger than the bear and the bear was defeated. The bear asked to keep on playing and fighting a few more times. They did, and the hikuli again toppled the bear. The bear sat on a rock and cried. Then he left and never came back."

It's quite possible that the free migration of black bears between Mexico and the western United States has aided their survival despite the threats that existed for them throughout their troubled history. Recently, DNA studies have found that bears from the Rocky Mountains in Arizona and those from the border between Mexico and Arizona share the same lineage. As long as the Mexico and United States border remains clear of fences, black bears will continue to roam the northern reaches of Mexico.

Black bears are true omnivores, feasting daily on salmon, sedge grasses, berries, crab apples, and skunk cabbage to round out their diet.

"LIKE CLOCKWORK,
BEARS APPEAR FROM
THE FOREST EDGE,
TAKING ADVANTAGE OF
THE DROPPING TIDE,
AND AN ANCIENT
RELATIONSHIP BETWEEN
HERRING AND
BEARS UNFOLDS."

—IAN MCALLISTER

GREAT MONARCH OF THE COASTAL WILDERNESS

IAN MCALLISTER

In the world of bears and people, seldom is the ocean recognized as a life-support system for the great monarchs and icons of North America's wilderness. To the north, because of their known association with the ocean, polar bears are afforded protection under the Marine Mammal Protection Act. Yet here, along the wet rain-forest-covered archipelago and mainland river valleys of Canada's west coast, a diverse population of bears also looks to the ocean for its survival.

For 20 years, I have followed spirit bears—as well as black bears and grizzlies—as they emerge from their high-elevation, cedar-dominated den sites—through the seasons of spring and summer, up until the shadow of winter when the great bears turn their backs on the dying days of the salmon runs. I have been continually impressed with the elegant relationship between these coastal bears and their ocean home. And each year, the defining line between these two worlds becomes less clear.

Just a few miles from where I live, humpback whales, white-sided dolphins, and countless sea lions follow vast schools of herring, a small forage fish so important to the many predators they serve that the fish is considered a "foundation" species on the coast. In March, the small fish, traveling in biomass measured in the thousands of tons, migrate to their natal intertidal spawning grounds, and when conditions are right, miles and miles of intertidal shoreline become covered in tiny white eggs stuck to rock and seaweed.

And like clockwork, bears appear from the forest edge, taking advantage of the dropping tide, and an ancient relationship between herring and bears unfolds. The ocean and rain forest are once again inseparable. The thick layers of tiny eggs become an easily accessible and abundant food source. Then, after about three weeks of feasting on eggs, other intertidal delicacies such as clams, mussels, barnacles, and limpets become the bears' focus. This seafood is a well-deserved and nutritional food source after a long winter's sleep.

By May, the bears move inland to harvest berries and other understory delicacies, and some move even farther toward the large alluvial floodplains that support estuaries rich in Lyngby's sedge, a plant rich in crude protein and a mainstay for coastal bears and dozens of other species.

Ecologists have been studying the contribution that the ocean brings to forest productivity, in particular the role that salmon play in fertilizing the forest with their decaying bodies. By tracing the molecular signatures of marine-based nitrogen in trees and plants, scientists are able to estimate the contribution that salmon bring to land. It is possible to find traces of salmon within the boughs of a 250-foot-tall spruce tree. And bears are given credit for facilitating much of this upslope nutrient transfer by carrying thousands upon thousands of carcasses into the forest where their partially eaten remains end up in the forest soil.

The Great Bear Rainforest is home to a globally unique population of bears, one that supports two genetically distinct black bear populations that diverged from their continental counterparts about 350,000 years ago. It is possible to see black bears, spirit bears, and grizzly bears all in a single day in a single river system. This coastal rain forest is truly a bear paradise.

But this bear kingdom, and the ocean world that they depend upon, is now under threat. The Great Bear Rainforest, having escaped the industrialization of much of the rest of North America, now finds itself precariously placed in the middle of plans to ship liquid natural gas and Alberta tar sands oil to Asian markets via pipelines and supertankers. The thousand-mile-long twin pipelines would cross the Rocky Mountains and the Coast Mountains ranges, whereby they would feed an endless procession of tankers headed to markets across the Pacific. But before they get to the open ocean, they will have to weave through an archipelago of islands so rich in unique genetic diversity that it is often referred to as the Galápagos of the north.

The inevitable risk of an oil spill would be felt by the bears on countless levels. Their intertidal food supply of barnacles, mussels, clams, and countless species of plants would all be contaminated. The life cycle of salmon, so intertwined to the rise and fall of tide and the flow of fresh water, would also be severely disrupted. The problem for these bears is that if their main food supply is disrupted they do not have other options.

British Columbia supports one of the greatest diversity of bears in the world; however, bears continue to be treated as an expendable resource. Each spring and fall, black and grizzly bears are legally hunted for sport and trophy. This hunt simply does not make economic, scientific, or ethical sense. Why kill the second slowest reproducing land mammal, next to the polar bear, for a trophy?

Bears are sentient, intelligent animals and they deserve a quality of life. I don't presume to know exactly what that means to a bear, but it surely does not mean having their rain forest home put at risk due to oil spills. I do not believe that we can evolve as a caring society when we allow animals to be killed out of greed or ignorance. A federally legislated ban on tankers through the fragile waters of the Great Bear would ensure that this bear kingdom by the sea is allowed to endure.

British Columbia needs to position itself as a forward-thinking society, one that is caring and respectful of the animals that we share this beautiful province with. In the end, if we knowingly place our most iconic land mammals in direct harm, what does that tell us about ourselves as people? We still have an opportunity to protect these globally important populations of bears from such a risk. It is with hope for the future that we do.

In the Great Bear Rainforest, bears use all levels of their habitat—from the intertidal zone, estuaries, and rivers to the forests and mountains *(opposite)*. In order to protect bears, we must protect entire ecosystems.

preceding pages

The first time I saw a spirit bear, it appeared on my camera trap. That inspired me to work longer hours, endure more rain, and wait more patiently for that rare gift in nature when I would see my first bear in the flesh. *(page 158)*

For more than a week, pilot Steven Garman and I soared over the rain forest in his Cessna 185. Little did I know that images from our scouting flight would provide an evocative representation of the terrain we would live in for two fall seasons in search of the elusive spirit bear. *(pages 160–161)*

This is the moment I met the most amazing bear. After I had spent months coming up empty, this large male appeared and stayed with me for two days, allowing me to live a childhood dream of walking through the forest with one of the planet's rarest animals. *(pages 166–167)*

A mother searches for Pacific crab apples while her black cubs wait patiently for a chance to nurse. I waited an entire season for spirit bears to come down to the rivers to feed on salmon, but their preference seemed to be crab apples and berries.

A female spirit bear navigates the dense rain forest looking for a crab apple meal *(opposite)*. The harvest was so bountiful that she didn't bother fishing for salmon.

following pages

In a previously logged forest, this ghostlike spirit bear chose to rest under one of the last old-growth forest patches in the area. This was to be one of my greatest days in the company of bears.

Without any dominant bears in this area of the river, this youngster takes its time to scratch away its itches from head to toe.

It was a magical moment when I put away my long lenses and photographed this bear up close in the rain forest, capturing intimate images of a great animal *(opposite)*.

following pages

A spirit bear approaches the river. I waited 12 days for ten-plus hours a day for this young bear to emerge out of the forest. The sense of place brought about by these moments is what I dream about as a photographer.

The rivers of the Great Bear Rainforest teem with migrating and spawning pink salmon. Many creeks are overfished each season, draining the rivers of the life-sustaining food that nourishes many species.

A bear, indifferent to my presence, carries a pink salmon in its mouth just feet away from me. It is these intimate images that transport people into the bears' world and make them care.

A spirit bear eviscerates a pink salmon, revealing the desirable egg sac *(above and opposite)*. After slicing the fish's belly, it relished each and every egg before moving on to the next fish.

following pages

When the bears are hungry, they devour every last bit of the fish, including the tail.

A petroglyph in the Great Bear Rainforest reminds us of the First Nations people, who for millennia have watched over these bears. Today, with First Nation guides like Marven Robinson, spirit bears and the Great Bear Rainforest have guardians and protectors.

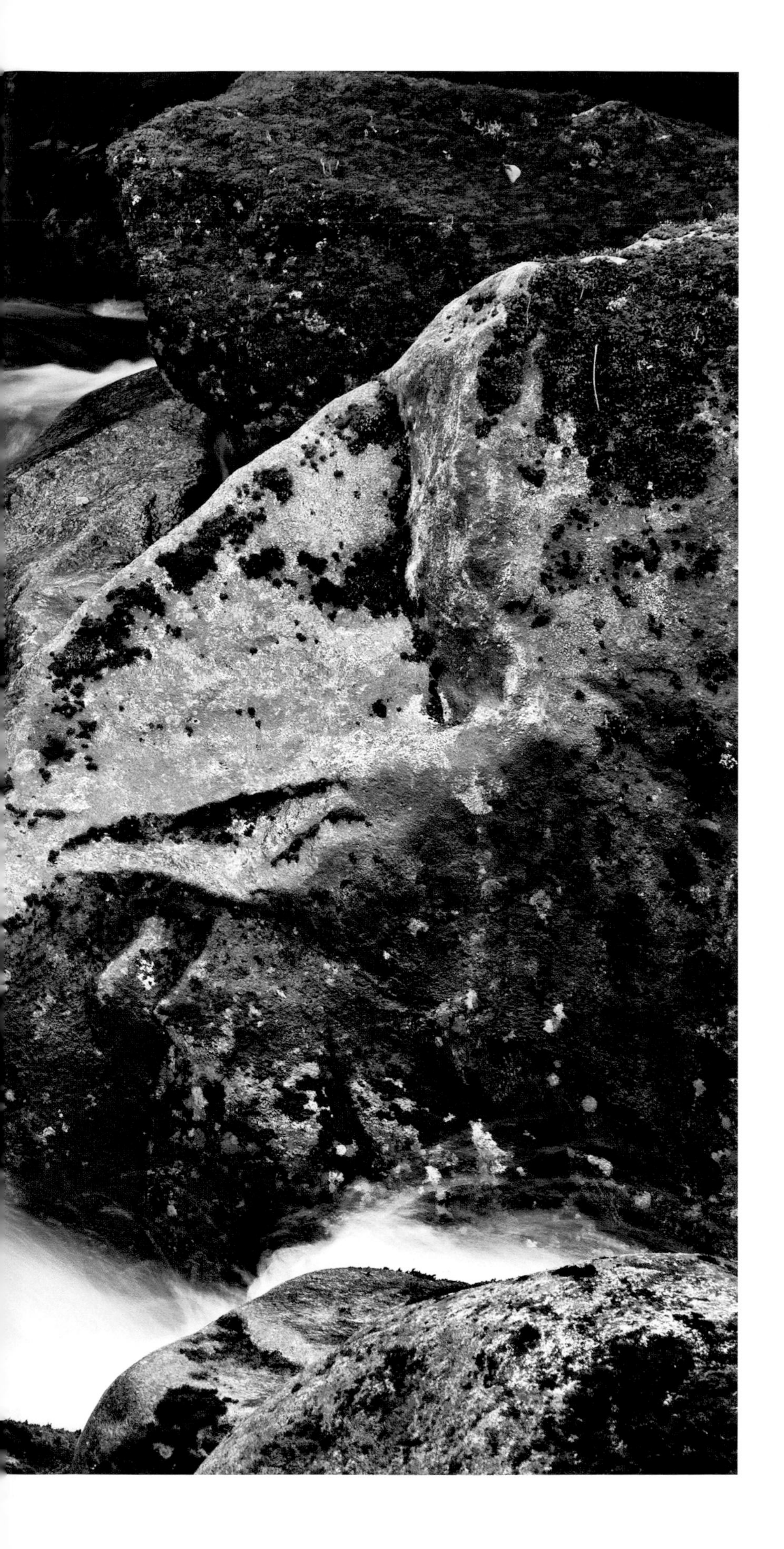

A lone spirit bear waits patiently for salmon to swim by *(opposite)*. I shot hundreds of long-exposure images of this scene, waiting for the bear to stop moving its head.

following pages

Mist envelops the Great Bear Rainforest, making it a very dark and wet place to work; even so, this moisture provides the water and energy for the thirsty forest and steep streams.

A PERSONAL PERSPECTIVE

LEGEND OF THE GHOST BEAR

MARVEN ROBINSON

My people, the Gitga'at First Nation of British Columbia, call him the ghost bear or Moksgm'ol. The spirit bear that roams the Great Bear Rainforest is a black bear with a white coat—a unique genetic trait that occurs in one out of every ten bears. When the white settlers arrived here they thought it was just a legend. For our people, who are a part of the T'simshian First Nation, Moksgm'ol has always had a very special meaning. I remember when I was growing up in the Gitga'at village of Hartley Bay, near Princess Royal Island, British Columbia. When someone saw a white bear in the forest, we were told not to talk about it. We didn't want it to be disturbed; we didn't want people to know about it.

The legend of the spirit bear is one that has strong roots in our mythology. When the land was green, our creator, the Raven, decided to give us a reminder of when the world was covered with ice and snow. He asked the black bear to turn every tenth offspring into a white cub. Raven promised that they would have unique powers. They would lead chosen people to special places and have the ability to find fish, deep in the ocean. Raven set aside a rain forest home for Moksgm'ol on Princess Royal Island, where he would live in peace and harmony with the T'simshian people forever.

The spirit bear shares a complex relationship with the old-growth temperate rain forest that we call home. It touches and interacts with every level of the ecosystem. From the salmon streams to the estuaries to the meadows, the bears rely on an intact wilderness to survive, and in turn, they help feed the land by depositing salmon carcasses deep into the forest. In this way, nutrients are moved from one part of the forest to another.

Our people also depend on the gifts that the forest and the ocean give us for our survival. For me, guiding tourists to see the spirit bear in the summer, when it feeds in the salmon streams, has turned into a livelihood. I am lucky to have found a way to share my passion for this animal with other people. I believe that I have a special relationship with the spirit bear, and in the 20 years I have spent tracking bears in these forests, I have never carried a gun and there has never been a bear attack on any tourist.

The relationship between the forest, the First Nations, and the spirit bear is indeed a sacred one. But today the peace and harmony the Raven promised are threatened by industrial forces beyond our control. If the Enbridge Pipeline, which will bring oil from the Alberta tar sands to our coast, is built, we run the risk of having a catastrophic oil spill in our waters. This would forever change the fate of these forests, of the spirit bear, and of my people. Our entire future is at stake, and it gives me great pride to know that all the other First Nations on the coast feel the same way. Perhaps now, by sharing our spirit bear with the rest of the world, we can help change the fate of our coastline.

A satiated large male spirit bear takes a nap under an old-growth tree in the Great Bear Rainforest.

A polar bear travels across the icy landscape after a dip in the cold polar waters. Loss of the life-sustaining ice is threatening the existence of this icon of the Arctic.

EPILOGUE

KARINE AND ANNA HAUSER

To roam the last corners of Earth where wild creatures still thrive is a privilege reserved to only a select handful. But even though we may never feel the chill of Arctic air through the frozen flap of an icy tent or we may wish that we were somewhere warm instead of chasing bears across the tundra, we understand where Paul Nicklen's passion comes from. It is the same fire that fuels the urgency so many of us feel to make sure that bears, and indeed all wild creatures, have a place to be wild. With his camera, Paul has taken us to some of the most remote areas of the planet, to places where wild bears still roam; and in our journey through these pages we have emerged gifted with a larger view of our planet.

In an ideal world, there would be species and places that are held sacred, kept untouched, where wild things are allowed to be wild and where humans only go to feed their spirit. These untouched, sacred places would become our biological capital to help sustain life on Earth.

This book is about building a greater awareness of the responsibility of what it means to be a human; it is about understanding that the history of every living thing that has ever existed on this planet also lives within us. It is about the ethical imperative—the urgent reminder that we are inextricably linked to all other species on this planet and that we have the duty to act as the keepers of our fellow life-forms.

Idealism aside, there are pragmatic actions that must be taken in order to tackle the issues that affect bears. Ultimately, however, everything relies on cultural perception and value. And by value we do not necessarily mean financial value. A dead bear on the wrong side of a rifle scope certainly has some value, but many would argue that a living bear has an intrinsic value that is hundreds if not thousands of times higher, and whose disappearance would be an ecological tragedy. Is it not true that a living bear represents a living forest—a carefully fine-tuned system that provides fresh water, clean air, pollination services, weather regulation, and myriad other benefits to towns, cities, and individuals? Many issues in modern society are rooted in economics, but there are those sacred values that demand that the pieces of the puzzle that allow life on Earth to exist be treated with reverence. It is impossible to put a dollar price on a system that supports us all. So why then are we undervaluing the very fabric of a healthy ecosystem?

Our system of values needs to change and especially the way we regard nature, wilderness, and its inhabitants. Bears and other species should be prized more alive than dead or in inadequate captive environments. This should in turn permeate our decision-making process at all levels. It is time for a cultural shift in our relationship with nature in our everyday lives: in the media, education, and throughout all aspects of society. Initiatives promoting the development of human consciousness and its identity toward a respectful approach to our environment and its inhabitants must be encouraged widely.

In our ideal vision of the world, taking into account our ecological and environmental impact will be one of the pillars of our value system, one that is not open to debate but simply a truism of how the world works. We will all recognize that as a species we rely on biodiversity, on balanced ecosystems, and on the Earth's limited resources for our survival, and therefore our impact on those must never be anything other than sustainable.

Global conversations are taking place every day now. While we can lament the fact that our current consumer habits have the power to destroy forests we will never see, we can also be hopeful. Globalization is our very own Janus. While it has the power to destroy, it also has the power to save. It creates new paradigms, allowing us to think of the world as one ecosystem, of which we are an integral part. It can take us one step further than those iconic photographs of the Earth from space, and allow us to take part in these stories for the next generation, and to meet other actors involved in the same unfolding drama on the other side of the world.

So what needs to be done to create a blueprint for a global future that includes bears?

The issue of deforestation and loss of habitat sits at the top of the agenda, as it not only displaces bears and other apex predators but also unravels the entire web of life, and more important, releases enormous amounts of carbon into the atmosphere. Nations need to come together to protect large areas of wilderness so that they become our natural capital for the future. As individuals, we must respect the intrinsic long-term value of these places. We are richer every time we choose to protect rather than destroy, as this wealth, once lost, cannot be replaced.

As an iconic symbol of beauty, innocence, and power for many cultures around the world, bears should be held in the highest regard. It is time to end the cruel practices that exploit them and are driving them to extinction. Stopping the extraction of bear body parts for food or as ingredients for traditional medicine or luxury beauty products requires a massive effort to educate those who sell and consume these goods. We also need to pursue legislation against trophy hunting of bears. How are we to learn to respect our living planet and ecosystems as long as killing large predators for sport is aspirational?

We need to aim for higher values of respect and compassion for all wild creatures. A bear in its natural habitat, exhibiting natural behaviors, makes for a much more awe-inspiring viewing than the sight of a predator, bear or otherwise, performing in a circus. Let our fascination focus on the animals themselves rather than on our attempts at dominating them.

While some bear populations are endangered, others are still prospering. We must keep a watchful eye as species like the Asiatic black bear become increasingly rare and the demand for bear products shifts toward bear populations that are thriving.

There is much good we can aspire to. Ensuring best husbandry practices for animals held in captivity is one such aspiration. Also lofty is the desire to lessen the impact we have on the wild spaces where bears and other wildlife live. Shielding our protected areas from the pressures of humanity is imperative. Wildlife, as much as people, requires peace and respite

So what needs to be done to create a blueprint for a global future that includes bears?

The issue of deforestation and loss of habitat sits at the top of the agenda, as it not only displaces bears and other apex predators but also unravels the entire web of life, and more important, releases enormous amounts of carbon into the atmosphere. Nations need to come together to protect large areas of wilderness so that they become our natural capital for the future. As individuals, we must respect the intrinsic long-term value of these places. We are richer every time we choose to protect rather than destroy, as this wealth, once lost, cannot be replaced.

As an iconic symbol of beauty, innocence, and power for many cultures around the world, bears should be held in the highest regard. It is time to end the cruel practices that exploit them and are driving them to extinction. Stopping the extraction of bear body parts for food or as ingredients for traditional medicine or luxury beauty products requires a massive effort to educate those who sell and consume these goods. We also need to pursue legislation against trophy hunting of bears. How are we to learn to respect our living planet and ecosystems as long as killing large predators for sport is aspirational?

We need to aim for higher values of respect and compassion for all wild creatures. A bear in its natural habitat, exhibiting natural behaviors, makes for a much more awe-inspiring viewing than the sight of a predator, bear or otherwise, performing in a circus. Let our fascination focus on the animals themselves rather than on our attempts at dominating them.

While some bear populations are endangered, others are still prospering. We must keep a watchful eye as species like the Asiatic black bear become increasingly rare and the demand for bear products shifts toward bear populations that are thriving.

There is much good we can aspire to. Ensuring best husbandry practices for animals held in captivity is one such aspiration. Also lofty is the desire to lessen the impact we have on the wild spaces where bears and other wildlife live. Shielding our protected areas from the pressures of humanity is imperative. Wildlife, as much as people, requires peace and respite from the modern world. We must aspire to reconnect with our primitive selves in the world's remaining wild spaces.

As we learn more about our relationship with wild nature, we must aspire to shift from a mind-set in which the destruction of an apex predator is acceptable to one in which the creature is not only respected but also revered.

We need to articulate the desire for a culture in which people are engaged in the protection and care of our future prospects; this small shift can have enormous consequences for a more sustainable planet.

If we want the children of our children to live in a world where bears still roam wild and free, we need to help our ailing planet heal. Without biodiversity so beautifully represented by iconic species such as bears, life on Earth as we know it today will not endure.

Hauser Bears, a charitable organization committed to the conservation of bears worldwide, focuses on education and research, putting emphasis on changing attitudes to effect lasting changes for all bear species.

ACKNOWLEDGMENTS

This book is a tribute to all of the great bears that have allowed me into their world. They are ambassadors of their dwindling habitats, and I am humbled and honored for having the opportunity to get a special glimpse into their secretive lives.

I want to thank the hundreds of people with whom I share the urgency to understand and protect wild ecosystems so that these great bears can continue to roam free.

The photographs collected here are a record of my ongoing journey into the wild. I may have been the one behind the camera, but I was fortunate to have many inspiring travelers with me along the way, not least the National Geographic team who carried this book to fruition: Susan Blair, Sanaa Akkach, Barbara Payne, Bill O'Donnell, and Barbara Brownell Grogan.

The vast majority of the pictures in this book are a result of the incredible assignments, support, and belief from *National Geographic* magazine. My sincere gratitude to Kathy Moran, who has been a friend and who has guided me through the process at National Geographic on many projects over the past 12 years. To Elizabeth Krist, Ken Geiger, David Griffin, Kent Kobersteen, Kurt Mutchler, John Q. Griffin, Chris Johns, John Fahey, M. J. Jacobsen, Maura Mulvihill, Declan Moore, Terry Adamson, and Stacy Gold for believing in my journey.

To the National Geographic photo engineers who continue to mesmerize me with their creations to help us get new and innovative images in the field: Dave Mathews, Walter Boggs, and Kenji Yamaguchi. Kenji, your patience knows no bounds.

An unquantifiable thank-you to my wonderful guides for their strength and for sharing their lifetime of knowledge and skill while in the field: Andrew Taqtu, Danny Taqtu, Roland Taqtu, Dexter Koonoo, Adam Qanatsiaq, Gideon Qaunaq, Samson Ejangiaq, Pakak Qammaniq, and Olayuk Niqatarvik. In memory of my friends Karl Erik Wilhelmsen, Seeglook Akeeagook, and Timut Qamukaq, who were all tragically claimed by the land they loved so much. I will never forget the laughter and priceless experiences we shared in nature. To Shaun Powell, Jed Weingarten, and Nathan Williamson, who stood by my side during fair and foul weather, and never wavered in the face of risk.

To my high school friend Brian Knutsen, who safely and skillfully flew our ultralight airplane around Baffin Island. To my dear friend Mike Gill, for sharing his wealth of knowledge about the struggling polar regions. To my pal Goran Ehlme—the "Swedish Cooler"—for keeping me warm in the polar elements in his award-winning dry suits by Waterproof. To Dr. Mitchell Taylor, who shared his knowledge and allowed me to accompany him on many scientific expeditions into the polar regions.

Bear
Paul Nicklen

Published by the National Geographic Society
John M. Fahey, *Chairman of the Board and Chief Executive Officer*
Declan Moore, *Executive Vice President; President, Publishing and Travel*
Melina Gerosa Bellows, *Executive Vice President; Chief Creative Officer, Books, Kids, and Family*

Prepared by the Book Division
Hector Sierra, *Senior Vice President and General Manager*
Janet Goldstein, *Senior Vice President and Editorial Director*
Jonathan Halling, *Design Director, Books and Children's Publishing*
Marianne Koszorus, *Design Director, Books*
R. Gary Colbert, *Production Director*
Jennifer A. Thornton, *Director of Managing Editorial*
Susan S. Blair, *Director of Photography*
Meredith C. Wilcox, *Director, Administration and Rights Clearance*

Staff for This Book
Barbara Payne, *Editor*
Sanaa Akkach, *Art Director*
Cristina Mittermeier, *Assistant Editor & Contributing Writer*
Carl Mehler, *Director of Maps*
Marshall Kiker, *Associate Managing Editor*
Judith Klein, *Production Editor*
Galen Young, *Rights Clearance Specialist*
Katie Olsen, *Production Design Assistant*

Manufacturing and Quality Management
Phillip L. Schlosser, *Senior Vice President*
Chris Brown, *Vice President, NG Book Manufacturing*
George Bounelis, *Vice President, Production Services*
Nicole Elliott, *Manager*
Rahsaan Jackson, *Imaging*

The National Geographic Society is one of the world's largest nonprofit scientific and educational organizations. Founded in 1888 to "increase and diffuse geographic knowledge," the Society's mission is to inspire people to care about the planet. It reaches more than 400 million people worldwide each month through its official journal, *National Geographic,* and other magazines; National Geographic Channel; television documentaries; music; radio; films; books; DVDs; maps; exhibitions; live events; school publishing programs; interactive media; and merchandise. National Geographic has funded more than 10,000 scientific research, conservation and exploration projects and supports an education program promoting geographic literacy. For more information, visit www.nationalgeographic.com.

For more information, please call 1-800-NGS LINE (647-5463) or write to the following address:

National Geographic Society
1145 17th Street N.W.
Washington, D.C. 20036-4688 U.S.A.

For information about special discounts for bulk purchases, please contact National Geographic Books Special Sales: ngspecsales@ngs.org

For rights or permissions inquiries, please contact National Geographic Books Subsidiary Rights: ngbookrights@ngs.org

978-1-4262-1176-8

Printed in China

13/RRDS/1